INDUSTRY STANDARD

V2003

VERSION 2.0 AUGUST 2003

THE READER IS ASSUMED TO HAVE THOROUGH KNOWLEDGE OF OEE AND RELATED ISSUES.

OEE Foundation

Edited by Arno Koch
Published by Makigami BV
Brabantlaan 6
5735 KA Aarle Rixtel
The Netherlands

@ info@oeeFoundation.org

ⓘ www.oeeFoundation.org

Content

INDUSTRY
STANDARD

Preface

"Just scrape the barrel" was the title of a cover story about organizations published in "Intermediair", a professional Dutch journal read by many managers. "We don't need to pump more money into it, we should work more efficiently", is what several politicians said. "Productivity can only be increased by making massive investments into ICT", is another frequently heard statement. "For reasons of efficiency, maintenance is being outsourced", is another favorite saying.

When you look for a deeper meaning behind these types of slogans, you will first encounter an enormous confusion of concepts. What is actually meant by efficiency, effectiveness, and productivity? And is it indeed true that improvements in this area always involve large investments, or is that precisely not the right thing to do?

Efficiency, effectiveness, and productivity ....

What is the relationship between *efficiency*, *effectiveness* and *productivity* and what is the path that can be followed to bring about 'improvement'.

Efficiency is determined by the amount of time, money, and energy – i.e. resources – that are necessary to obtain certain results. In order to meet our daily production quota, we commit a specific machine that uses up energy, make operators and maintenance personnel available, and provide raw materials. For example, if we are able to meet our daily production with less energy and fewer operators, we have operated more efficiently.

Effectiveness is determined by comparing what a process or installation can produce with what they actually produce; therefore, effectiveness does not tell anything about the efficiency – the amount of resources that have to be committed to obtain that output. If we are successful in manufacturing more good product in the same time period, effectiveness will increase.

Productivity is determined by looking at the production obtained (effectiveness) versus the invested effort in order to achieve the result (efficiency); in other words, if we can achieve more with less effort, productivity increases.

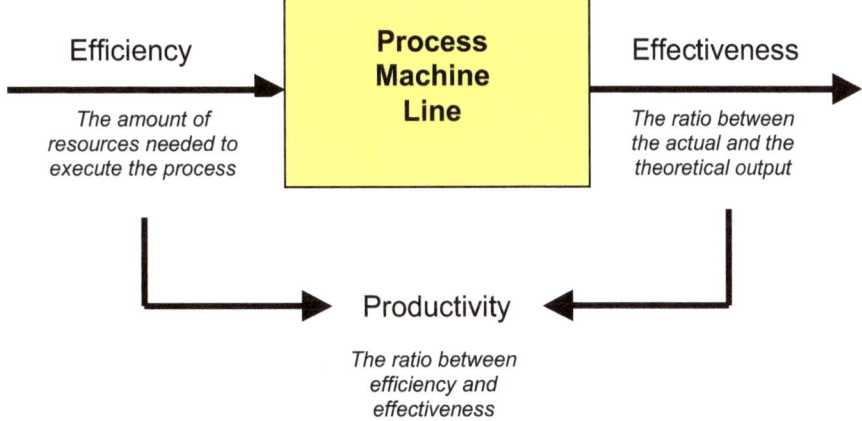

Goldrath ('The Goal') defines productivity as: *'the extent in which a company generates money'*. The goal of a production company is therefore not to reduce expenses but to generate as much money as possible!

Is improvement still possible?

Most striking when looking at the traditional approach to improvement, is that focus is often exclusively on efficiency; the famous cheese slicer continuous to slice production further and further.

How much room for improvement is still left at the input (efficiency) side? 10%? 20%? And does it still make sense to try to reduce another operator or engineer, or to put pressure on the buyers to negotiate even more competitive prices?

As is often the case, that question can not directly be answered. If the supplier can give us a better price because we help him with managing his production process better or – as we see in the automotive-industry – we force the supplier to go through basic improvement processes, such as Lean Manufacturing or TPM, not only the price will decrease, but the quality and the reliability in delivery will increase as well. That is good news for both parties.

However, by managing solely by keeping the cost price down, you run a large risk of saving pennies per product but losing many euros or dollars due to stoppages, quality losses, etc.; in other words, 'penny wise, pound foolish'. Many production teams can give you striking examples of that.

Increasing output

We strangely look less often at the output side – the effectiveness – of the equipment. Apparently, the output is more or less considered to be 'as it is'. However, every line manager knows that the installation will spontaneously start to run better simply by standing beside it and giving it attention. When you check the logbooks, there are days that, on occasion, the installation produced spectacular amounts of good output.

It happened to go well that day...

Ask the team how that happened and you will hear a precise run-down of all elements that went right that day. The raw materials arrived on time and were of the correct quality, the installation kept on running and was set correctly, the right people were present, it was not too warm, etc., etc. This is often regarded as a fluke and nobody is wondering how you could create a similar situation a second time. That is strange actually, for if it can happen once, why should it not be possible to happen again. And if it can happen a second time, why not always? Usually, a whole series of "Yes, but's" will follow...

Suppose you would write down those "Yes, but's" and turn them into a list of action items. What would that give us?
To be able to answer that question, we will have to dive a little deeper into the world of Effectiveness.

Our machines run non-stop!

What determines the effectiveness of an installation? First of all we must address the question of whether it does or does not run. Roughly there can be three reasons why an installation is not running:

- The installation quit; it broke down.
- The installation could be running technically speaking, but is waiting for something; materials, an operator, filling, to be set-up, etc.
- The installation could be running, but is not planned in because there is more capacity than demand.

Of course, the ideal machine would never break down and would never have to wait for anything; therefore, it would be running all the time as long as there is demand for the product.

Our machines run at top speed!

Subsequently, the effectiveness is determined by the speed at which the installation is running. This is always a tricky topic, for what is the maximum speed? The speed at which it is at the verge of breaking down? Or the speed at which the quality of the output reaches the bottom limit of its spec? This will guarantee a lively discussion. It is useful to see that it is often simply unknown what the maximum speed is, while the maximum speed that people come up with is usually based on various assumptions (which, in turn, you could turn into an interesting list of action points!). An example of such assumption is: *"If I ran the machine any faster and the material got jammed, we would suffer major damage"*. Why does it get jammed? Is that always the case? When not? What must happen to prevent it from jamming again? Why does damage occur when it gets jammed? What do you have to do in order to.... etc.

We have 'zero defects'!

If the actual speed is determined versus the theoretical speed, the next effectiveness determining factor can be looked at: Does the realized output meet the set quality standards? It can be quite an eye opener, if you ask ten different people on the shop floor to indicate very clearly when a product does or does not meet the specs, you will receive ten different answers. It becomes even worse when it turns out that the one who produces the product, the operator, cannot determine, or cannot unequivocally determine this. Also here lie many opportunities to solve all "Yes but's" and to ensure that the person who makes the product is also able to determine whether he is manufacturing a good product, so that he can keep the quality within pre-set specifications.

Are we 'ideal'?

Thus, if there is demand, an ideal and effective machine is always running at maximum speed without producing any out-of-spec products. We can assume this to be 100% effective. We know that 100% effectiveness is impossible over a longer period of time; after all, installations must sometimes be maintained and converted. The guideline is that 85% is a realistic "World Class" value for "traditional" machines. That implies that the installation, for example, produces 99% of the products "First Time Right within Specs", operates at a speed of 95% of the theoretical maximum speed, and is actually running 90% of the operating time (99% quality x 95% speed x 90% running time = 85% effectiveness).

INDUSTRY
STANDARD

In a three-shift system it means that the installation runs for 90% x 24 hrs = 21:36 hrs at 95% speed with 99% quality. Consequently, there will be 03:24 hrs available for maintenance, conversion, and other possible waiting times. Incidentally, the 85% mentioned is a rather conservative figure; nowadays we see in the automotive-industry equipment that runs over 90%.

The analysis of hundreds of installations for various processes shows that, as rule of thumb, an average installation in an average (non-TPM) company runs at an effectiveness rate between 35 and 45%. Of course, there are always cases that stick out; for example, values in the pharmaceutical industry may lie considerably lower and there are also cases that show considerably higher values.

If it turns out that an installation has an effectiveness of 40%, while people always thought that there were limited options left for potential improvements, it is extremely good news: this means that twice as much good product can be manufactured (your effectiveness rate would be 80%!) at the present cost level. Or, you manufacture the same product with one shift instead of two.

Yes, but then the costs will increase!

It is often assumed that achieving such improvements will necessitate an enormous increase in costs for, for instance, maintenance. That is sometimes partly true, for example, when it concerns overdue maintenance and you are then actually paying off a loan, because a fundamental design flaw has to be solved (and, therefore, you can also see this as paying off a postponed cost item). However, by activating the knowledge that is present on the shop floor in the right way, 80% of the improvements can often be implemented without any capital expenditures and at minimal costs.

It is not so hard to imagine that an installation, which halts on a regular basis for various reasons, or whose process is not stable enough to operate at high speed without any losses in quality, automatically requires more resources at the input side as well! Reversely, it may be that lowering the efficiency (for instance, by spending a little bit more money and time on preventive maintenance) will bring about a strong increase in effectiveness, which - bottom line - creates a higher net productivity. Such considerations can only be made if, in addition to efficiency, particular attention is paid to losses in effectiveness as well.

In all cases, it is necessary to take decisions concerning actions leading to improvement on the basis of facts and figures describing the entire productivity picture.

World Class Manufacturing does not accept any losses at all. That is what management must focus on and management must have the will to go further than mere window dressing and scratching the surface. Unfortunately, that is often even harder than just opening the wallet. Companies that do take this route, discover over and over again: There still lies a nearly unlimited potential for improvement for those who learn to see it and seize it!

Arno Koch

The History of the OEE Industry Standard

Working as a sr. Lean consultant I noticed two things over and over again:

1. Operators, line managers and management all either want to compare equipment, or are afraid that this will be done.
2. Setting up the definitions for gathering OEE data brings up the same discussions over and over again. Larger companies all seem to struggle -most political- fights about how OEE is defined, one wants to include PM, the other takes it out, and the third takes it out under conditions... thus giving fear on the shop floor that 'unfair comparisons' are done.

Ad 1. Although I feel OEE is a shop floor tool, not meant to benchmark, it is certainly possible to use certain elements as useful reference information considering it is done in the right way.

Ad 2. For every problem there is only one optimal solution. So why discover the wheel again...

So I started to wonder if it would be possible to define a kind of an 'Industry Standard OEE definition', that would make sure that at least within the same company everybody uses OEE in the same way. i.e. if we are talking about 'Availability' at least it should be clear that everybody in- or excludes the same issues (i.e. breaks, PM, etc).

What I did was this:
*   I took ALL OEE registrations I had ever seen (quite some...) and figured out where the common denominators were.
*   I grouped all possible OEE elements in a logical way.
*   I tried to give all 'negotiable' elements a clear definition.
*   I validated every group as 'Production', 'Failure', 'Idle' or 'Unscheduled'.

At first I thought this would become a huge document since the equipment I looked at was varying from refineries to cement- and paper mills to beer breweries, food processors, drilling, stamping, welding, plating.... well you name it and is was there.
Guess what... I figured out the whole lot of it fits on a couple of A4!

Then I took this concept to some experienced OEE implementers and discussed every element. Every time we had a discussion, I gave all arguments I had heard and tried to find the Best Of Best argument, considering it had to be applicable on ANY other situation! In fact this process is still going on, but the picture becomes quite clear.

It is my objective to have an OEE standard definition available where every choice is companioned with very reasonable and strong argumentation, that (if I did my job well) can not be refuted within the spirit of **TPM** and **Lean Manufacturing**.

Now, in 2003 many parties have joined us on this adventure and tried hard to 'doubt' every element of the standard; yet the arguments given by the first group of forum-members stand firm. However we have gathered some additions and clarifications which are added to this version of the OEE Industry Standard.

I now want to invite every OEE using company to join us in this standard:
1. **to check if the definition can stand up to new discussions;**
2. **to get broader support within the industry, so we get more unity in OEE definitions.**

If you would like to contribute to the discussion please let us know by sending an email to: **info@oeeFoundation.org**.

Arno Koch

Section 1: Scope

Several parties, such as production teams, line managers and top management may have a different scope when looking at 'effectiveness'. Being aware of those differences, it is possible to calculate different indices representing those different scopes; **all based upon the same data**. In the literature we find several attempts to do so, unfortunately they are not always consistent.

Although this definition goes beyond the scope of OEE, it is necessary to have a clear picture on this issue since it determines how to define certain categories within the OEE.

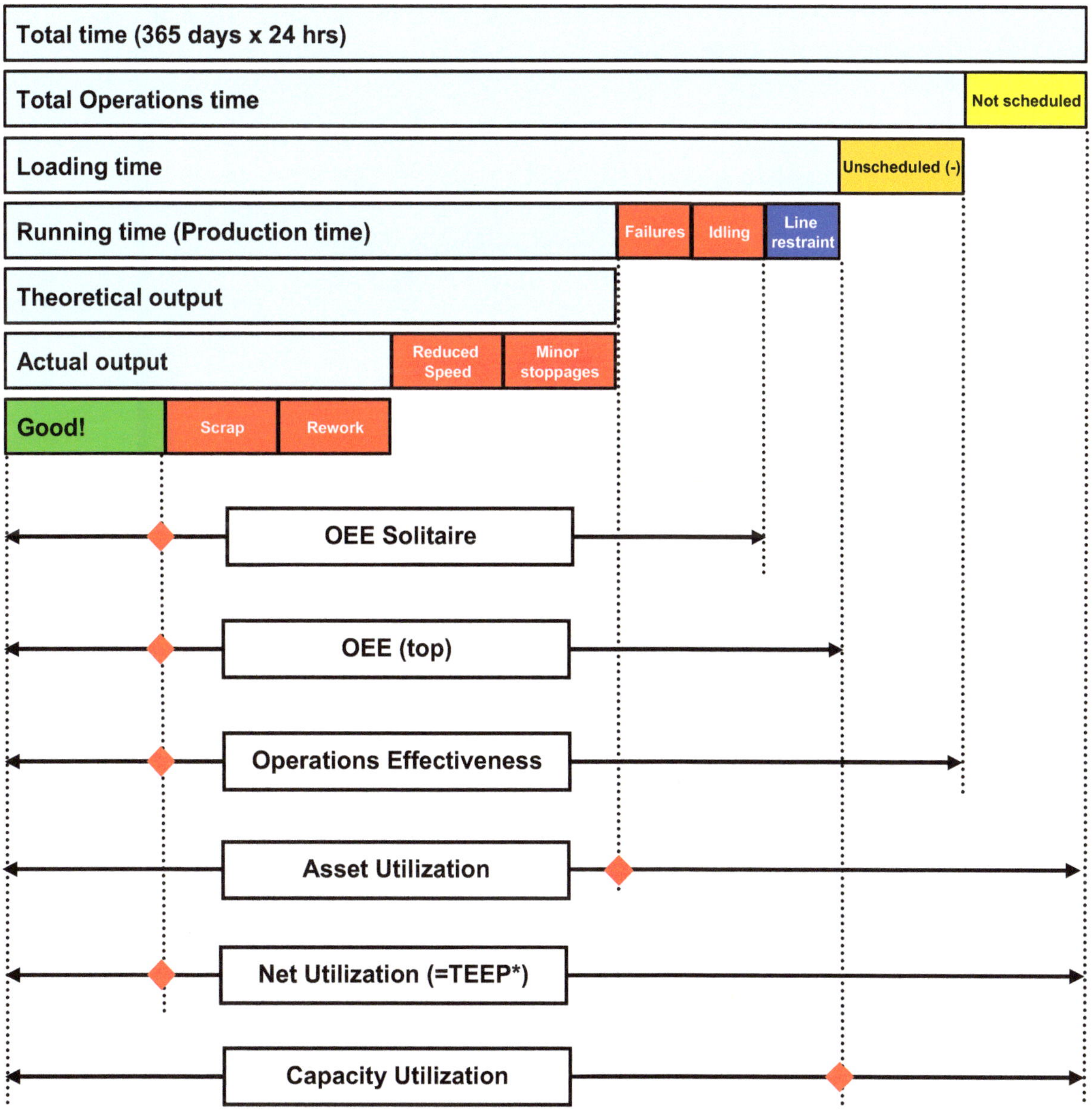

Total time (365 days x 24 hrs)	
Total Operations time	Not scheduled
Loading time	Unscheduled (-)
Running time (Production time)	Failures · Idling · Line restraint
Theoretical output	
Actual output	Reduced Speed · Minor stoppages
Good! Scrap · Rework	

- OEE Solitaire
- OEE (top)
- Operations Effectiveness
- Asset Utilization
- Net Utilization (=TEEP*)
- Capacity Utilization

1.1 Brief description of different scopes

1.1.1 OEE Solitaire

OEE Solitaire is only applicable in situations where the monitored equipment has a technical link to other equipment/processes. Thus the equipment is an integrated part of a line. The effectiveness of such equipment is partially depending on factors like line balance and effectiveness of other equipment. 'No Input' (input buffer empty) and 'No Output' (Output buffer full) are the exponents of such influences.
In certain case i.e. with huge installations in process industry parties may want to know the effectiveness of the equipment independently from the rest of the process. In other words; What would be the effectiveness of the equipment when it was running solitaire, with no restraints from the rest of the line?

For this purpose, an OEE can be calculated excluding 'No Input' and 'No Output' thus showing the effectiveness purely focussed on the equipment independent from the line.
Calculating the OEE Solitaire takes Line restraint time (normally considered to be (I) Idle time) out of the loading time (so it is considered as Unscheduled time (-)).

Caution: 'No Input'/'No Output' should not be confused with for example 'no raw material' or 'no packaging' since these have an other cause; they are not Line Restraints!

1.1.2 OEE

OEE is the default scope for a production team.
It shows the ratio between the theoretical maximum good output during the loading time vs the actual good output.
The loading time can be less then the operations time since the equipment can be unscheduled during the operations time, thus reducing the loading time. Loading time therefore is the time the equipment was supposed to be running

In cases where several products have been produced, (either sequential or parallel) the Performance part of the OEE is calculated as a weighted average between the several expected output ranges.

1.1.3 OEE Top

OEE Top is equal to OEE except for the way the performance rate is calculated. In OEE Top the performance rate is calculated based upon the Name Plate capacity of the equipment; thus ignoring restraints due to the product-machine combination. It is used to detect effectiveness losses due to the chosen product mix.

1.1.4 Operations Effectiveness

Operations effectiveness goes beyond the scope of the production team (Maintenance – Operator – Engineer). It includes the operations time the equipment is not available to the production team or operations management does not want or can not schedule the equipment. Examples are [No Orders], [Legal Restrictions] like energy contracts and mandatory holidays and test runs for R&D.

1.1.5 Asset Utilization

The percentage of the total (calendar) time that the equipment runs.

1.1.6 Net Utilization (=TEEP)

Net utilisation reflects the ultimate improvement potential; it is the ratio between the total (calendar) time and the actual effective production time (or, if you wish; the theoretical maximum amount of good product versus the actual amount of good product coming out during that time). In some publications you may find the same definition named TEEP; Total Effective Equipment Performance.

1.1.7 Capacity Utilization (=Gross Utilization)

Capacity Utilisation reflects the ratio between the total (calendar) time and the loading time. It shows the hidden operations potential i.e. the difference between 3 and 4 shifts.

Section 2:   Definition of several types of time

Type	Name	Definition
P	**Production time = Running time**	Something is coming out of the equipment, regardless of the amount, speed or quality. A machine, spinning at top speed, not generating output therefore is not running…. Caution; batch producing equipment (like cookers, brick-ovens) are considered to run during their batch processing time.
F	**Failure time = Breakdown time**	The machine is not having output due to a machine related technical problem.
I	**Idle time = Waiting time**	The machine is not producing output because it has to wait for something (like a setup, or raw material) Therefore, waiting for a technician during a breakdown is not Failure time but Idle time!
L	**Line restraint time**	The equipment can not generate output because it gets no input from the line, or can not give its output to the line Applicable to equipment with a technical link to the line (pipe, conveyor)
—	**Unscheduled**	The equipment is scheduled out of the total operations time for reasons beyond the scope of the production team.
	Not Scheduled	Time where there are no operations-activities going on at all. 'The factory is closed, the lights are out'.

A machine is running when there is output,

regardless of the quantity or quality

INDUSTRY STANDARD

2.1 Groups and types of Time usages

2.1.1 Production time

No	Time Usage	Max. recommended	Time type				

Refer to Section 2 for explanation!

No	Time Usage		Max. recommended	P	F	I	L	-
1	**Production**	*Try to define no more categories!*	5	P	F	I	L	-

	Plating Stamping Assembly Filling Moulding Drilling Refining Baking	'Something is coming out of the machine' regardless the speed and its quality. In batch producing equipment: From the start of the *batch processing time* until the end of *the processing time*. While loading or unloading the equipment, the product is not being processed thus the equipment is NOT running!

2	**Reworking**	3	P	F	I	L	-

	Re-filling Re-pressing Remanufacture	The equipment is running, (re-)processing product that was not produced first time right. *Although this is normal production time in terms of OEE, in some cases it is useful to split the different types of production runs to make them visible in the pareto of time-usage.*

3	**Reduced Speed Run**	3	P	F	I	L	-

	Running ½ die Reduced Setup Tact	The equipment is running at reduced speed; the equipment output is deliberately slowed down to balance a line or to execute an inline setup. *Although this is normal production time in terms of OEE, in some cases it is useful to split the different types of production runs to make them visible in the pareto of time-usage.*

2.1.2 Failure time

No	Time Usage	Max. recommended	Time type				
10	**Failure [function x]**	10	P	**F**	I	L	-
	Failure filler Failure Capper Failure depalletiser	A failure (=breakdown) prevents production due to a technical reason in the machine. The categories should be chosen process oriented and should describe the phenomenon, not the cause. *Choosing functional categories (like Electric, Mechanical, Pneumatic) is not supporting the cross-functional production team efforts.*					
11	**(Re)Adjustment**	2	P	**F**	I	L	-
	Adjustment Re-calibration	The machine is stopped WHILE A BATCH RUNS because adjustment is needed, to keep/get a product in spec. It is seen as a failure because at a certain point the equipment is no longer capable of producing the right spec due to a technical malfunction that needs to be corrected. The process is not capable of a stable operation. *Adjustment and calibration after setup belong to startup.*					

Time categories should be PROCESS related

rather than FUNCTION related

2.1.3 Idle time

No	Time Usage	Max. recommended	Time type				
20	**Setup**	5	P	F	**I**	L	-
	Product change Setup Quality Change	A setup in terms of OEE lasts from the last product A until the first product B, *regardless its quality*. (remind, this is different from the SMED definition: Last GOOD product until first GOOD product!) *Depending the situation it might be useful to precede or succeed the setup time by a shutdown and startup category, to make visible in a pareto where the main losses of setup are to be eliminated.*					
21	**Startup/Shutdown**	2	P	F	**I**	L	-
	Heat up Start after stop Run-in Checking machine Preparation Pressurise Pump dry Empty out Clear out Rinse out Cool down Calibration	The machine is not producing due to the fact it has no vertical startup or shutdown. This can occur at beginning or ending of shift but also before or after a setup or repair works. This time is defined separately from the main reason to shut down the machine. *Cleaning belongs here if it is part of the startup or shutdown activity, including startup of a new product (process-cleaning).*					

OEE
INDUSTRY
STANDARD

No	Time Usage	Max. recommended	Time type				
22	**No Operator at machine**	5	P	F	**I**	L	-
	Break Meeting Training Help at other machine 'Washing hands'	The machine is available, but is not running because the operator is not operating it for example during his break, a training or a meeting. Just like Preventive Maintenance, this time needs to have a balance between doing enough but not too much.					
23	**Quality problems - Process disturbance**	2	P	F	**I**	L	-
	Stopped because output is out of spec	The machine can not run because the process can not meet up with the specifications for (at that moment) unknown reasons. *If the reason is known (i.e. bad raw material), it should be registered as such (Bad raw material = Waiting for correct material).*					

No	Time Usage	Max. recommended	Time type				
			P	F	I	L	-

24	**Refill/Replace**	3	P	F	**I**	L	-
	Blade change Refill ink cartridge Oil Refill	The machine is stopped while refilling or replacing auxiliary, e.g. cutting blades, coolants, ink, oil, etc.					

25	**Loading**	3	P	F	**I**	L	-
	Load labels Load boxes Load Raw material	The machine is stopped while refilling/loading raw material, e.g. packaging materials, foils, and other raw materials.					

26	**Handling**	2	P	F	**I**	L	-
	Truck (un)loading Forklift driving Container change	The machine is waiting while needed material is being handled. *This is a more specific situation from the 'Waiting' block.*					

27	**Waiting**	7	P	F	**I**	L	-
	Waiting on Tooling Waiting on Engineering Waiting on Maintenance Waiting on Inspection Looking for Gauge No/Bad Packaging No/Bad Raw material	The machine is not running because, for example, the correct raw material or a tool is not present at the moment it is needed, or an action can not be performed. *In this block we find reasons for idling that are **caused by a weak planning**. In the block 'No Resources and No Personnel' the reasons are 'Force Majeure'.*					
	No operator Available	It might occur we forgot to schedule or replace an operator.					

No	Time Usage	Max. recommended	Time type				
			P	F	I	L	-

28	**Autonomous Maintenance & Cleaning**	3	P	F	**I**	L	-
	Shift Maintenance Daily Cleaning	The main reason why the machine was stopped was due to AM activities, including cleaning. If a machine is cleaned in the time it was standing due to another reason (ie waiting for raw material) the real reason why the machine is standing is being registered. *Process cleaning (like rinsing between two products) belongs to startup-shutdown.*					

29	**Preventive Maintenance during Loading Time**	2	P	F	**I**	L	-
	Planned PM during Loading time	Scheduled time to perform PM.					
	Unplanned PM	The machine is stopped at a not scheduled moment to perform PM or… The machine is stopped at a scheduled moment but exceeding the scheduled time to perform PM during a scheduled time. *PM happens 'in process'; it is scheduled in the production sequence and needs to have a balance between enough and not too much. If well performed, the planned idling time will reduce the unplanned breakdown time.*					

OEE
INDUSTRY
STANDARD

2.1.4 Line restraint time

No	Time Usage	Max. recommended	Time type				
50	**No Input (in Line process)**	5	P	F	I	**L**	-
	Empty input buffer No input due to pre-heater No input due to de-palletiser	The machine stands still due to no input from a preceding process that has a technical link, e.g. a conveyer. *In lines it can be useful to get an overview in the pareto to show what process decreases the effectiveness from this machine the most. In that case the time usage is split in two or more items like;* *[no input due to process x].*					

No	Time Usage	Max. recommended	Time type				
51	**No Output (in Line process)**	5	P	F	I	**L**	-
	Output buffer full Stop at palletiser Stop at tray packer Conveyer blocked	The machine stands still due to the fact it can not get rid of its output; mostly due to a stop in a succeeding process that has a technical link, e.g. a conveyer. *Also see block [No Input].*					

Remark:

The Line Restraint categories are used to reveal unbalanced capacities and/or timing in line processes. A line can be thought of literally: several machines tied together with conveyors, pipes etc.

However, applying Lean-Principles, those categories can be very well used to detect un-balanced situations between machines not technically tied together, e.g. in a cell or between several departments.

It even might be used to detect flaws in the supply-chain.

The [L] category therefore can be seen as a strict [L]ine Restraint but also as a [L]ean or [L]ogistic restraint.

2.1.5 Unscheduled time

During the operations time, the equipment may be unscheduled because:
- the product is not needed (over-capacity);
- it is not allowed to run (due to Governmental regulations or contracts);
- 'Force Majeur' (catastrophe's outside of the company);
- the equipment is 'handed over' to an other party (e.g. R&D).

No	Time Usage	Max. recommended	Time type				
60	**No Orders**	3	P	F	I	L	**-**

No Orders Over-capacity PM during No Orders Cleaning during No Orders	The machine is not required to run due to a lack of **customer** orders; the capacity is not needed. (customer = the one who **pays** for the product!) *Beware!* A warehouse will never be 'the customer'! **Do not hide over-production!**

No	Time Usage	Max. recommended	Time type				
61	**No Personnel Available**	1	P	F	I	L	**-**

No Personnel Strike	The machine is not scheduled for production because there is no operator present due to 'force majeure' like • Strikes • Massive influenza • Poor availability of workforce in the region. *If there is no operator available due to a lack of planning (so the machine was supposed to run), choose 'no operator at machine' or 'wait for operator' (both [I]dle).*

No	Time Usage	Max. recommended	Time type				
62	**No Resources**	2	P	F	I	L	**-**

River Frozen Energy Contract Boycott	The machine is not scheduled for production because a resource is not available due to reasons OUTSIDE of the company, including energy; thus reasons other then lack of planning or handling (in such cases: Waiting).

OEE
INDUSTRY
STANDARD

No	Time Usage	Max. recommended	Time type				
63	**Test Production**	1	P	F	I	L	-
	R&D Test Run	The machine is not scheduled for production to make it available for product development, testing new products or processes etc.					

INDUSTRY
STANDARD

2.1.6 Idle Time registered as Unscheduled Time

Activities that normally would have been performed during loading time, can sometimes be scheduled outside loading time. In a two-shift operation, the machine might be setup or cleaned during the night or in the weekend, in absence of the regular crew. Preventive Maintenance could be carried out at a moment no production was scheduled. In a 3 shift operation from Monday to Friday, the PM could be scheduled in the weekend, thus not affecting the loading time.

In this way OEE may be increased by using non-loading time for activities that normally would affect loading time.

Doing so has some serious disadvantages:

- The activity is necessary to perform the scheduled production thus can not be skipped: This means the OEE is artificially high. When Loading Time needs to be extended (e.g. more capacity is needed) and 2 shift becomes 3 shift, or 3 becomes 5 shift, those activities automatically will fall into the loading time, thus dropping the OEE!

- Usually, outside the loading time, there is no pressure to get the machine *up and running*; this might result in a less effective performance of the activity scheduled outside of the loading time.

It is advised to register ALL machine related activities, necessary to perform the operation, WITHIN the loading time, regardless of the normal shift-time. Thus PM on Saturday would be Idle time and decreases the OEE!

However, if you choose not to do so, make sure to register ALL the time the machine is being touched or activated outside normal loading time. In this way the potential loss can be identified and made visible.

No	Time Usage	Max. recommended	Time type				
			P	F	I	L	-
70	**Preventive Maintenance outside Loading Time**	2	P	F	**I**	L	-
	Planned PM Outside Loading Time (Annual) Overhaul	PM necessary to maintain a high availability of the equipment, being performed outside of the normal Loading Time. In a 3 shift operation from Monday to Friday, the PM could be scheduled in the weekend, thus not affecting the regular loading time. *Overhaul is a specific form of preventive maintenance. If the machine is unscheduled for a longer period of time to be completely overhauled, it should be considered [PM during loading time] thus (I)dle. Only if the overhaul is performed at a moment the equipment would anyway not be running (e.g. because the facility is closed) it belongs in this block. If it is done in a period where there are no orders, [No Orders] is the reason for stopping the machine, and not the overhaul. Thus [No Orders] is registered! To identify PM during No Orders, make a separate category: [PM during No Orders].*					
	Unplanned PM Outside Loading Time	PM outside loading time can take longer than originally planned; this can be made visible here. If PM in Loading Time takes more time than planned and is being continued outside Loading Time it becomes [Unplanned PM Outside Loading Time].					

No	Time Usage	Max. recommended	Time type				
			P	F	I	L	-
71	**Activities performed outside Loading Time**	3	P	F	**I**	L	-
	Cleaning Outside Loading Time Setup Outside Loading Time Loading Outside Loading Time	Activities that normally would have been performed during loading time, performed outside loading time. In a two-shift operation, the machine might be setup or cleaned during the night or in the weekend, in absence of the regular crew.					

Maximum Time categories per machine +/- 20;

All categories should fit on one side of the OEE registration form!

INDUSTRY
STANDARD

Section 3:   Speed definitions

To calculate the performance rate, the **theoretical maximum speed** is related to the actual speed.

3.1   Time versus Quantity

The maximum speed can be calculated in time or in number of product;
i.e.
- maximum speed is 10 seconds per product
- maximum speed is 6 products per minute

Nevertheless since OEE is primarily a shopfloor tool, and people on the shopfloor rather talk about 'units being produced' than 'seconds it has taken to produce a unit', it is preferred to register speed in units, not in time.

The parameters for Performance can be expressed in either

TIME or QUANTITY, but QUANTITY is preferred

3.2   Maximum speed of what?

The maximum speed can be determined from the NPC, the Name Plate Capacity (Design speed). However, mostly there are several product(groups) on a machine with their own derived maximum speed. In such cases, special care has to be taken not to include hidden losses in the maximum speed due to 'difficult products' which might refer to 'products which we don't control the process for'

3.3   Name Plate Capacity (NPC)

The NPC should not be taken granted for. More than once it was discovered to include all kind of hidden losses. The manufacturer might have chosen a low NPC for reasons of liability or to 'fit' the equipment with other equipment in a line.

OEE
INDUSTRY
STANDARD

3.4   The Standard

The derivation of different maximum speeds for different products should be done in a mathematical manner.

Example:
An extruder is designed (and checked for!) extruding max. 500 kg plastic per hour.
Product A consists of 250 grams plastic, product B consists of 500 grams.
Thus, the theoretical maximum speed for A is 500kg/250gram= 2000 pcs per hour.
The theoretical maximum speed for B is 500kg/500gram= 1000 pcs per hour.

The theoretical maximum speed for a product-machine combination is called 'The Standard'. It is specifically not called Norm, since this word has negative associations related to piece work for many people in different countries.

INDUSTRY
STANDARD

3.5 How maximum is maximum speed?

100% OEE =
The theoretical maximum capacity of the equipment

This statement should be taken serious. If not done so, situations may (and do) occur where the shopfloor is filled with scrap, the machine is suffering one breakdown after the other and still accounts for 80% or more OEE. How?

70% Availability, 80 % Quality, 143% Performance = 80% OEE !

As soon as the performance rate goes over 100% (indicating the standard is chosen too low!) the beautiful balance of the OEE parameters is broken, and the focus may be taken away from what it is all about: **identifying and reducing losses**.

In cases where the maximum speed has to be determined based upon a Best Of Best analysis, it should be considered that this BOB is achieved under the former and current circumstances, including current losses.
Since in the end even standards are broken by product- and equipment improvement, the BOB should not be considered too easily as maximum value. As a rule of thumb the BOB value should be raised with at least 10 to 25% to serve as Standard.

A well chosen Standard will only change
when the product or the machine fundamentally changes

Section 4:   Quality definitions

Producing 'quality' means producing **a product that meets its specification,** not by trying more than once but **First Time Right.**
Products not meeting its full spec, but still useful and possible to sell (i.e. as B product, or in a different market) are not first time right and thus should be considered 'scrap'.

If a product does not meet its spec, but can be reworked, in terms of OEE it is to be considered as 'scrap', but can be identified as a special form of scrap, by labelling it as 'rework'.

> In terms of OEE, Scrap, rework and sub-spec are the same:
>
> It was 'not first time right'; therefore it is a loss

- Defining scrap-product may reveal poor specifications or poorly testable specifications!
- Good specifications always refer to the needs of the customer!

INDUSTRY
STANDARD

Section 5: Discussion issues

Q: **Why do you include breaks in the OEE? We have a legal right to have a break!**

A: This approach assumes the machine can not run while you are having your coffee or lunch. The reason to include it in the OEE is to make the production team aware of this potential loss. Is there really no way to let the machine run 15 or 30 minutes without operator interference? Would it be possible to have an other operator at the machine? Could an other operator watch your machine while you go away?

Q: **We feel cleaning and maintenance should not be included in OEE. This is necessary to keep the machine running well!**

A: Exactly! So Maintaining a machine is not meant to reduce its effectiveness... No, we clean and maintain to RAISE its effectiveness. By taking this time out of the OEE, we will never see if the effort we spent to clean and maintain is bringing us a higher effectiveness at the bottom line!

Q: **Ok, but at least you should take out Preventive Maintenance. I, the operator, do not have any influence on preventive maintenance!**

A: You, the operator, are part of a production team. Together with your maintenance and engineering colleagues you are responsible for the effectiveness of the equipment. If you, the operator, can prove with facts and figures that you are suffering severe losses due to repeated breakdowns and too little (preventive) maintenance, or due to too frequent PM, it serves a mutual interest if you bring this up in your production team meeting!

Here too: **If PM reduces your effectiveness: stop it, otherwise: do it!**

Q: **You want us to track a maximum of 10 failure categories, but I want to define 85. How else can I ever know what bold is breaking?**

A: At first glance that seams to make sense. But OEE is not a breakdown registration system, but a loss detection system.
Let me explain the consequences: Imagine after 3 months registering OEE data, it shows the main loss is in availability. So now you want to know what you need to do to get your availability up. So you take one of the seven tools and draw a pareto diagram of all your time-events. Lets assume each of the 85 breakdown items occurred at least once. What you will see is a pareto with an immense long tail, not giving a clear clue where the main losses are located.
In the other approach where you would have registered failures on, lets say, 5 process parts of the equipment (like 'incoming conveyer, pre-heater, moulder, compressor, outgoing conveyer') it would show what part of the equipment is restraining the process most. Then you start temporarily to focus on that part. A simple registration card (maybe even showing the 85 breakdown items) for some weeks will give detailed insight what's going on. Sometimes it is a matter of simple maintenance, sometimes a Small Group Activity can solve the problem for once and forever. By this circle of focused improvement the equipment will become better and better. This example also shows another disadvantage of the 'breakdown registration

INDUSTRY STANDARD

approach'; if you handle each breakdown in the best way, thus eliminating it forever (either by reengineering or taking it into PM), it will not reoccur, while other breakdowns may start to occur. So after a while the breakdown registration system will not reflect the reality anymore. In the other approach this is less likely to happen.

Q: If we have no raw material, or we wait for a technician, we start to clean. What do we register?

A: The reason why the equipment stopped running is not the cleaning but the 'waiting for technician' or the raw material. Always register the true reason why the machine is stopped, not how you spent that time.

If you would like to add a discussion item, please let us know: info@oeeFoundation.org.